Natural Man Or Earth Man And

Planetary Influence Over The Earth

And Its Inhabitants

Karl Anderson

CHAPTER XI

OF NATURAL MAN, OR EARTH MAN, AND EVOLUTION OF SPECIES, AND PLANETARY INFLUENCE OVER THE EARTH AND ITS INHABITANTS.

In Genesis, second chapter, commencing at the fourth verse, we find : —

" These are the generations of the heavens and of the earth when they were created, in the day that the Lord God made the earth and the heavens.

5. And every plant of the field *before it was in the* earth, and every herb of the field *before it grew:* for the Lord God had not caused it to rain upon the earth, *and there was not a man to till the ground.*

6. But there went up a *mist* from the earth, and watered the whole face of the ground.

7. And the Lord God formed man of the dust of the ground, and breathed into his nostrils the breath of life; and man became a *living* soul."

Now compare with this Gen. i. 26 : —

26. " And God said, Let us make man in our image, after our likeness : *and let them have dominion* over the fish of the sea, and over the fowl of the air, and over the cattle, and *over all the earth*, and over every creeping thing that creepeth upon the earth.

27. So God created man in his own image, in the image of God created he him ; *male and female created he them.*"

I have before described these men in the *image* of God to be the planet sof our solar system, male and female, and ordained to have dominion — see Gen. i. 26 above — *over all the earth :*

and at which time (see Gen. ii. 5) there was not a man
(or Adam). And these planets were also to have dominion
over everything movable upon the earth as well as the earth,
and especially "over every creeping thing that creepeth upon
the earth." It is well to take notice that in this last account,
although plants and herbs were already created, yet on this
earth there were none in it, nor as yet any herb grown; also
that there was as yet no rain, but a thick mist covering and
watering the whole earth, showing that the heat of the earth
was still insufferable and full of malarious exhalations, and fit
for no life as yet, and until rainfall it could not be. Now, it is
a well-known fact that as soon as the heated season commences
at the tropics, during morning and evening, the death-dealing
malarious fog or mist covers the whole face of the earth. This
hangs over the valleys and rivers and the trees and sides of the
mountains until dispersed by the sun and the land breeze; and
to sleep in it is almost certain death, and never does it fail to
produce terrible malarial fevers, chills, and dire disorders.
Rising like a sheet to a level of about ten feet, no dwellers in
the tropics can hope for any health unless they keep to upper
chambers. But after the rainy season commences, these death-
dealing mists or fogs are beaten down, and the air is less
noxious, and existence becomes pleasurable. A good description is thus given by Rider Haggard in "Allan Quatermain":—

If one desires health, the night air and early morning are to be avoided
in tropical climes, and the sun should be at least an hour high before one
should stir much about; and when the mists arise into clouds, and the glori-
ous sun asserts *his* sway, then one can enjoy life. Until then, —

"There is silence upon the face of the earth and the waters thereof;
Yea, the silence doth brood on the waters like a nesting bird;
The silence sleepeth also upon the bosom of the profound darkness.

Only high up in the great spaces star doth cry unto star.
The earth is faint with longing, *and wet with the tears of her desire.*
The star-girdled night doth embrace her, but still she is not comforted.
She is enshrouded in mists like a corpse in the grave clothes,
And stretches her pale hands to the east.
Lo! away in the farthest east there is the shadow of a light;
The earth seeth and lifts herself; she looks out from beneath the hollow of
    her hand.
Then thy great angels[1] fly forth from thy holy place, O Sun;
They shoot their fiery swords into the darkness and shrivel it up;
They climb up the heavens, and cast down the pale stars from their
    thrones;
Yea, they hurl the changeful stars back into the womb of the night;
They cause the moon to become wan as the face of a dying man —
And behold! *thy glory comes, O Sun!*"

It is plainly shown by the writer of Genesis that not only the period of caloric or heated cycle, when the earth was not yet cool enough to precipitate rain, and human life was not yet existent, was meant in this description, but also that a tropical region in a tropical climate was the first abode of man.

To digress: our Old Testaments are exceedingly incorrect in much of their rendering of the Hebrew, and therein lies much false reasoning occasioned by such errors. Probably one of the most important of omissions is that of the following : —

GEN. ii. 3: " And God blessed the seventh day, and sanctified it: because that in it he had rested from all his work which God created and made."

Now, a very important omission is here at the end; it is this: "which God created and made *to create by evolution,*" this omission being לעשות, or *laassass;* my authority for this being the Rev. Fleurlicht, doctor of divinity, and a

---

[1] In the above the rays of the sun, or angles of light, are designated as angels, which I have before explained.

celebrated Hebrew scholar. Without this it would be difficult
to understand the remainder of this second chapter of Genesis; but with the command to create by evolution, and the
putting away of the idea that man is coequal with his Creator
and not merely the highest order of animal creation, we can
get rid of a deal of self-conceit, and be willing to learn, and
not so ready to teach, matters about which one knows not even
the slightest rudiments.

By this we find that the fifth verse is correct, and that this
new-formed earth had naught of plants or herbs even, nor man
nor beast, naught but a ball of plastic mud, not yet cooled; that
all the other things which emanated from the Almighty Father
were created, but not made; and that they were already upon
the planets Saturn, Jupiter, Mars, sun, moon, Venus, and Mer-
cury, which were to have dominion over the earth and all things
therein. Thus the magnetic effect of these planets is at once
apparent. They do have dominion over the earth and all things
upon it.

GEN. ii. 7: "And the Lord God formed man of the dust of the ground
[thus saith the version at hand; in others it reads *slime*], and breathed into
his nostrils the breath of life; and man became a living soul."

Here is no mention of creation, but distinct making, form-
ing, an effort of physical power or of energy. He *formed* man.
No mention of this man being an image of himself; neither of
being bi-sexed or self-creating; merely, he formed him of the
dust of the ground, and breathed into his nostrils the breath
of life; and man became a living soul; *i.e.*, that as a soul or
ray of the Father he had existed from the beginning; and
now commencing with the lowest order of creation he must
evolve through every gradation up to the highest; and thus

endowed with a physical or earthly body, the incarnated soul became what is called a *living* soul, or man. And yet he had always lived, and will always live, since every death is but a change in the order of evolution. And so "ye must be born again." The original first man is called Adam. In the Hebrew, Adam means red earth (primitive soil).

The situation of the Garden of Eden is purely tropical, as the description of its boundaries shows; extending from Hindostan to the Nile in Egypt; a climate wherein grows "every tree that is pleasant to the sight, and good for food," and also where the aboriginal or first men had no necessity for clothes, and went naked; and only by migrating northward out of these latitudes they found colder and colder climates, and became toughened and savage from famine and want, and slew the beasts for food and used their skins for clothing. These thickly populated tropical regions prove conclusively that man's original dwelling-place is between the tropics.

GEN. ii. 18: "And the Lord God said, It is not good that the man should be alone; I will make an help meet for him.

19. And *out of the ground* the Lord God formed every beast of the field, and every foul of the air; and brought them unto Adam to see what he would call them: and whatsoever Adam [or man] called every living creature, that was the name thereof.

20. And Adam gave names to all cattle, and to the fowl of the air, and to every beast of the field; but for Adam [man] there was not found an help meet for him."

Therefore as yet man is but the highest order of animal creation. But there are none of this lower order through whom he can propagate his species; for nature despises a descent in scale, and always produces a monster or mule, incapable of any reproduction. So there is no beast or animal found that can propagate the highest species of animal, the man, unless the counterpart of man, the female.

GEN. ii. 21. "And the Lord God caused a deep sleep to fall upon Adam, and he slept: and he took one of his ribs, and closed up the flesh instead thereof;

22. And the rib, which the Lord God had taken from man, made he a [*womb*-man] woman [or female], and brought her unto the man."

25. "And they were both naked, the man and his wife, and were not ashamed."

The foregoing proves the ancient belief in evolution, also the original source of primitive man, and his happy state of ignorance and comparative purity, in a climate of perennial summer and never-failing fruits, fed by the bountiful hand of nature and continual harvest.

GEN. ii. 8: "And the Lord God planted a garden eastward in Eden; and there he put the man whom he had formed."

10. "And a river went out of Eden to water the garden, and from thence it was parted and became into four heads.

11. The name of the first is Pison: that is it which compasseth the whole land of Havilah [Hindostan]. . . ."

13. "And the name of the second river is Gihon: the same is it that compasseth the whole land of Ethiopia [this is the Nile, and the former the Ganges].

14. And the name of the third river is Hiddekel: that is it which goeth toward the east of Assyria [the Tigris]. And the fourth river is Euphrates.

15. And the Lord God took the man, and put him into the garden of Eden to dress it and to keep it."

According to this description, we include the entire region inhabited by the ancient Aryan and Hindu races, and their offshoots and co-worshippers of the sun, the Persians, Babylonians, and ancient Egyptians; and from the Chaldeans the Israelites obtained the true spiritual worship of God. For all the ancient priests were instructed in that one great truth, that God is one and indivisible; and though they perverted truth, and held the ignorant multitude to a multiplicity of gods, both male and female, it was done to multiply gifts and levy constant contributions, through superstition, to enrich themselves, regardless of the suffering and poverty of their miserable devotees. And so blind obedience to their instructions as the only ordained ministers of God is always inculcated, and knowledge and learning are looked upon with abhorrence by them.

GEN. ii. 17. "But of the tree of the knowledge of good and evil, thou shalt not eat of it: for in the day that thou eatest thereof thou shalt surely die."

In the commencement of the Bible are given this command, and curse if not obeyed; all written for the succeeding generations, and as a warning to those to come not to be too full of curiosity, and then as a punishment for the so-styled failure of Adam and Eve to comply with this command, written down by onè whom no living person can name, as events happening at a time when only one ignorant man and woman existed on the whole face of the globe, who, standing in a tropical climate, named every animal and every living thing from the torrid and temperate, Arctic and Antarctic regions, and whose names no one ever remembered; but in every nation extant (and it seems that there were other populated places at the time Adam and Eve were created) these animals and living things were called by different names. Plainly the entire description is allegorical, and came also from a very primitive idea of how first man was produced. To find out the first cause mankind has never as yet failed to exercise all there is of scientific research; and to-day the greatest intellect has arrived no nearer the end than the earliest searchers after the secrets of life from immemorial ages in the past.

Knowledge, study, and scientific research soon do away with superstition and priestcraft, and exalt their possessors above all earthly cares.

GEN. iii. 22 : '' And the Lord God said, Behold, the man is become as one of us, to know good and evil : and now, lest he put forth his hand, and take also of the tree of life, and eat, and live for ever :

23. Therefore the Lord God sent him forth from the garden of Eden, to till the ground from whence he was taken.''

Plainly spoken this. As knowledge would enlighten man, he is driven back into laborious employment where he would have no opportunity to advance, and merely grub along to live, and become a slave to the soil, — become the *soodra* of the Hindus, or the *fellah* of the Egyptians, or the lowest caste among all the rest. It also alludes to regions outside the tropics where winter, spring, summer, and fall alternate, and where to live the soil must be cultivated, irrigated, and planted, and the harvest gathered for the winter months ; wherein is a constant struggle for life, brought upon man from his desire to wander, and his insatiable curiosity and dissatisfaction with a peaceable and quiet life, and perpetual desire for knowledge which causes him to seek even in the bowels of the earth after it ; and also the unsuccessful attempts of the ancients, and even those of present times, after the " elixir of life."   The help meet of Adam was called Eve. Eve signifies a serpent, on account of its fecundity.   She is called the mother of all living.

This Adam and Eve are the modern Joseph and Virgin Mary. For Adam is the sign ♌ (Leo), and Eve the sign ♍, or original Virgo Scorpio or serpent woman, the mother of all, or help meet of Adam, and signifies that the first degree, or the head of the virgin or Virgo, is lying and *meeting* with the last degree of the sign ♌, or the strong man or *sun* of God (*ruler of all things upon our earth*). Thus one sign continually *goes in unto* another sign, or they are joined like an unbroken circle or as *one* flesh. As the sun is in his greatest strength about Aug. 22, when just entering Virgo, and the seed begins to ripen into the future harvest, the Egyptians commemorated this gift of God in reproducing through the sowing of seed all that was good for life. And also, as the old seed which had been sowed had to become corrupt before the resurrection, or birth of the new, also because much seed was indigenous to the soil and sprang up as it were spontaneously and perennially, the earth, or ♍, was likened to a virgin bringing forth her fruits without the aid of man, and ruled over or o'ershadowed only by the Lord or sun. Therefore ♌ (Leo) and ♍ (Virgo) are conjoined. The Sphinx with the entire body of the lion and the head of the virgin bears testimony to Egyptian knowledge and to the astrological meaning of the virgin-born Saviour ; also to the fall of man, the story of Adam and Eve, of Ruth and Boaz, and the heat of the sun at the time of the year. In the constellation Leo (which you can easily find on any celestial chart, and also on the "Combination Horoscope" in this book) near the sign ♌ or the head of the lion, you will see a cluster of stars representing a man with a flowering staff in his hand, whose feet have the head of the virgin between them. This constellation Boötes, Boaz, or Adam was called by the Greeks Ἰοσεφ, or *Anglice* Joseph. At the feet of this figure, as before mentioned, is the celestial virgin or constellation Virgo, represented as a female winged figure holding a sheaf of wheat in one hand, and originally, before ♎ was introduced by the Greeks, her feet extended to the constellation Scorpio or serpent. So the serpent's head was bruised or crushed by woman, and he, the serpent, wounded her heel. Virgo by the Greeks was called Ceres, the bread-giver, or the harvest queen. So that the *earth*, or Adam's (red earth) wife, Eve, Ruth, Boaz, and the celestial virgin are one and the same, as also Ceres, and also the virgin mother of every crucified Saviour of the ten great religions.

In the third chapter of Genesis the account of the fall of man is a mixture both of phallus and planetary truths. The serpent, or tortoise, which is found as a symbol or emblem in all parts of the globe, — in China, in Egypt, among the Chaldeans, among many Indian tribes, and in the islands of the Pacific and Indian Oceans, and among the Mexicans or Aztecs, the Delawares and Mohicans or Algonquin tribes of North America, — is not only an allusion to the male organ of generation, but to the constellation Scorpio or original fall sign. For you must recollect that originally there were but ten signs of the zodiac, and that the sign ♍ (Virgo Scorpio) comprised the space now covered by ♍ (Virgo) and ♎ (Libra), which last is equivalent to Sept. 22, the time when nature commences to drop her fruits, or the sun (or Adam, as son of the sun) must be found wanting, or falls, or *fall* or autumn commences. As the ♏, the serpent, is ascending (represented as climbing a tree), — for ♏ in astrology rules over the organs of generation, both male and female, — the astrological meaning is that the sun (heat) having passed through Leo and Virgo, or the man and woman, or the earth or nature, and passed the summer months, and now going into fall, by reason of worms, decay, blight, etc., much fruit of the harvest is cast to the ground, also much unripe fruit eaten which causes cholera, cholera morbus, and many ills, and often death. It also signifies that the sun has reached the meridian, and that the hottest part of the day has passed and evening approaches, and sundown or fall below the horizon approaches. It also denotes the fall passover, when the sun reaches ♎ (Libra), a table, or feast of unleavened bread, where the sun leaving Virgo, a *garden*, or garden of Gethsemane, the +, or Christ, giveth the *Lord's supper*, and ♏, or Judas Iscariot, the betrayer of blood, who dippeth with him in the dish ; viz., ♎, ♏, conjoined. In fact, it is all *fall ;* for even in the mixture of phallus worship, in this sign the sun's number 6 has turned to the moon's number 9, and the ☉ falls in Libra, and Saturn (Satan) is exalted. And Libra is the house of Venus (♂). See the Exaltation and Dignities of the Planets. And here is the mystery of three in one.

Hence the foregoing embodies both phallus and planetary accounts ; and he or she who is enlightened or who has eyes that *will* see, may see clearly the fruit of the tree which is in the midst of the garden, viz., on the +, and the meaning of the so-called serpent, and of both man and wife partaking, and of the man being tempted by the womb-man ; also the angle (or angel) of the Lord which appeared at the annunciation ; for ♎ is a feminine sign, and is equivalent to 9, or the womb, or man's private parts *inverted.*

We also read that at first they were naked and were not ashamed, and that afterward they made them coverings of fig leaves.

Under the interpretation of the phallus worship or signification or symbolism, this is a fine representation of the heat and innocent state of youth until tempted by natural laws, and the inevitable feeling of shame attending the gratification of lust. They strove to hide it by covering themselves with the lightest clothing of a torrid zone, and the cooling leaves of the fig tree are chosen as a symbol; first, because the fig is full of seed; secondly, because the use of this slight covering is and always has been an aboriginal custom. During childhood, till the age of puberty, they go entirely naked, but as soon as betrothal takes place (and among savages this is in the very youth), and passion has been gratified, the circlet of the female is worn, and the slight covering of the man. The remainder of the body still in nakedness is thus astrologically explained. The season of summer is herein described. The trees formerly naked and bare are now clothed in luxuriant verdure, and the *fruit green* but apparent, is visible but unripe as yet, and not fully developed. Much falls to the ground and the earth, represented by the earthly sign ♍, the virgin, whose help meet is ♌, or Leo, or Adam and Eve conjoined. The virgin earth fecundated by the magnetic vital rays of the sun, *becomes as one*, for astrology teaches that wherever the ☉ and other planets and stars are in the heavens at a birth, these *are radically* the places during the life; and the magnetism of those planets, sun, moon, and stars, is fixed or as *one flesh or being*. The birth of Eve is a beautiful symbol of astrological teaching, for she, the moon, and Adam, the sun, must be in the same sign, in the due east or Garden of Eden, to find the serpent, which you will find in the *Mysteries of Isis* later on. In this entrance to Eden they must be joined *as one flesh*, or become as one body, and yet in the very entrance to Eden the head of the serpent appears. To resume: The virgin earth fecundated by the vitalizing magnetic rays of the sun, ripens the harvest, and yet under the early frosts of the forthcoming winter (for the sun has now got into Libra, and the serpent or Scorpio leads him, the sun, to his fall), and the superabundance of water, the menstrual of all nature, the heavy fruit, unripe and frost-killed, falls to the ground, and ripe only in appearance, the apples worm-eaten, another symbol of the serpent's work, are among other fruits thrown to the earth. To eat of these wind-falls and decayed and unripe fruits produces almost certain death; and yet not surely death in all cases. By the phallus explanations, lust satisfied, shame, disgust, and coldness ensue. Vitality is as it were dead, and yet not surely so. See the symbol 9 — in this case resurrection — He shall rise again.

Later on, as coldness ensues (*fall*) and winter approaches, which by the phallus signification is explained by the above-described coldness, and that the soul of man is now debased and animal;— the allegory of clothing them with skins of wild beasts signifies the age of puberty or hairy covering, by the astrological explanation. The sun is next approaching the winter solstice. The earth by frosts and cold has lost its fecundity, is cold and barren, the reeds and bushes and trees are denuded of verdure, and like hair cover the earth, and man who in summer has neglected to attend to his harvests for his winter supplies must labor hard to exist; and by not attending to his natural food — fruit and vegetables (if you read Genesis attentively you will see that he was to eat of every fruit while in Eden (the summer), and after the fall was to eat herbs or vegetables, but never at any time was there any mention of eating the flesh of beasts or even fowl or birds) — he has now to fight for a living, to slay the wild beasts and take their hides and hairy coverings to keep himself warm, the first recourse of primitive man. Eden is paradise to him no longer, for winter, cold, and frost have arrived, and man prodigal and wasteful has neglected his future and must wander like an outcast in search of food. He is driven out and a guard put at the east gate, or sign ♈, by order of the grand master, and out into darkness he passes, from the west gate (♎), thence down into the earthly sign of ♑ or into mid-winter, or driven into the barren earth or strong house of Saturn. And here the sun (Hiram), the grand master, whose stronghold is the lion (♌) of Judah, must be raised or resurrected; for ♑ (Capricorn) is "renewer of life." And here on Dec. 25 the sun again commences to climb up the ecliptic; the new sun is as it were born; the child of the + is to ascend into heaven after his death; the son of the sun is born; the child of ♍ becomes Emmanuel, or Lord with us; the son of the ☉ is hung between two thieves, ♈ and ♎; darkness is over the face of the earth. It is midnight. It is midwinter. It is under the earth, the descent into hades or hell. It is the house of the goat. It is the hour of midnight when "graveyards yawn." It is high twelve. It is the turning of the tarot or ⅃, or the phallus, ⅃, or o6go, from the positive noon and negative night, or ♋ and ♑, which two figures, 6 and 9, are signs of the favorite or strong house of Isis or the moon, viz., ♋, and which is the double serpent or the numbers of Ra and Isis, 6 and 9.

Thus is the world renewed by the representation of the birth of a child, thus personifying the sun's annual revolution. Thus the grand master performs his work, whom the three ruffians, ♈, ♋, ♎, have slain from the ⳥. Thus Masons meet,—and thus they part, L. And in winter where naught but the evergreen tree flourishes, in Capricorn, emblem of eternal life, here under the earth where one of the twelve fellow-craft (twelve signs of the Zodiac) stayed, you will find our grand master, who has lain in ♑ (Capricorn), a mausoleum, grave, or deep hole, for three days, and will now be raised again "amidst the *clouds* and mist of winter, in glory —*the light of the world."*   "And the Lord said, Let there be light, and there was light."

And this is called by theologians the original fall of man, an inevitable law of planetary movement ordained by the Lord of hosts from the beginning of the world, and which will never change, — a law which, explained either by phallus worship or planetary or astrological teaching, is but an allegorical description of the serpent wisdom, or knowledge of the *Gen-Isis*, goddess of nature, the sun on the ✕ between the cherubims, or *fixed signs*, the head of the bull, the head of the lion, the head of the eagle, and the head of a man, *fixed and unalterable* forever, and shows that God is not a man that one can reason with him, or that he alters.   It also denotes that if we disregard or abuse the laws of nature and pervert the spiritual exaltation of true love by the debasing influences of passion or lust, or the influence of Mars (♂), represented by its favorite stronghold, ♏, the serpent, then all that there is of Venus (♀) (represented by the *fall* sign ♎), of joy and happiness, is fled for evermore, and shame ensues.   From Eden expelled, and repelled from all that is good for man, the flaming sword, or remorse, meets them at every point, and he who originated spiritually from the sun, or Lord of hosts, is plunged downward to grovelling ideas of deceit and must purify himself through a series of re-births till he once more ascends in glory above the clouds which have darkened him, and he becomes the pure Ra, or purified ray of light ; for the natural law or phallus, or the obelisk, is but a shadow of the eternal law of the most high, the pyramid (see the combination zodiac), on which blazes in glory the eternal flame of fire, the image of the Creator.

CPSIA information can be obtained
at www.ICGtesting.com
Printed in the USA
BVHW011348120619
550844BV00007B/132/P